Marizângela Ribeiro dos Santos
Jane Geralda F. Santana
Cinara Soares P. Cafieiro

Chemistry teaching and everyday life in the education of citizens

Marizângela Ribeiro dos Santos
Jane Geralda F. Santana
Cinara Soares P. Cafieiro

Chemistry teaching and everyday life in the education of citizens

A peculiar look

ScienciaScripts

Imprint

Any brand names and product names mentioned in this book are subject to trademark, brand or patent protection and are trademarks or registered trademarks of their respective holders. The use of brand names, product names, common names, trade names, product descriptions etc. even without a particular marking in this work is in no way to be construed to mean that such names may be regarded as unrestricted in respect of trademark and brand protection legislation and could thus be used by anyone.

Cover image: www.ingimage.com

This book is a translation from the original published under ISBN 978-613-9-64212-0.

Publisher:
Sciencia Scripts
is a trademark of
Dodo Books Indian Ocean Ltd. and OmniScriptum S.R.L publishing group

120 High Road, East Finchley, London, N2 9ED, United Kingdom
Str. Armeneasca 28/1, office 1, Chisinau MD-2012, Republic of Moldova, Europe
Printed at: see last page
ISBN: 978-620-7-73492-4

SUMMARY

This study addresses one of the biggest challenges facing chemistry teaching in secondary schools, which is to build a bridge between school knowledge and the everyday world of students. From this perspective, chemistry education must focus on citizenship, which is also an education of the human conscience in terms of its ethical and moral values and the construction of a critical view of the world. These values need to be based on the principles of respect and equality, guaranteeing fundamental human rights in the face of commitment and duty to the new society. Based on these needs and the difficulties faced by educators, we carried out a bibliographical study to analyse the contribution of chemistry teaching to the civic education of students at the Federal Institute of Education, Science and Technology of Bahia - Guanambi Campus - BA. To develop this project, we opted for a qualitative-quantitative approach. At the end of the discussion and reflection stages and the analysis of materials, this research shows an Inefficiency in the teaching of Chemistry at the IF Baiano - Campus Guanambi in terms of its conception aimed at the critical and citizen training of its students, who initially characterise the teaching of contextualised Chemistry as simple exemplification and description of everyday facts or situations without the intention of teaching and understanding ideas of contextualisation such as addressing social issues, with a view to developing attitudes and values and transforming social reality.

Keywords: Citizenship; education; contextualisation and society

CHAPTER 1

INTRODUCTION

To understand that chemical thinking, i.e. the deliberate use of defined procedures to obtain certain products, was present even in primitive civilisations, extracting the metals contained in ores and using plants for medicinal purposes, is to understand the social relevance of chemistry for a society that is increasingly dependent on technological innovations and scientific advances.

Since then, chemistry teachers have been able to use this study of the development of the science of chemistry as a contributing factor in the formation of citizens who are aware of their role as educators in the face of social issues. The contribution of the chemistry educator is understood not to be the "transfer of knowledge", like "bank teaching", but the contextualisation of chemical knowledge with social aspects, relating chemical concepts to the students' daily lives and a reconstruction of ideas with the aim of enabling the student to unravel existing realities.

On the other hand, educators experience the realities of their students, learning from them. In other words, educators are no longer just educators, but those who, while educating, are educated. We mustn't forget, however, that we have a special role in the complex production of chemical knowledge. We are chemistry teachers, or rather chemistry educators, and in this sense our knowledge is of a special nature. Rather than 'advancing' specific chemical knowledge, we are committed to recreating it in the school environment and in the minds of the younger generations of humanity. Therefore, the teaching of chemistry needs to be contextualised, where the focus of knowledge is the conscious formation of the citizen.

Educating citizens doesn't just involve teaching the chemistry of polymers, polyamides, polycarbonates, hydrocarbons and sulphamides, as some textbooks make up for everyday life. The chemistry we need also involves developing ethical values. What is important is the need to reformulate the teaching/learning of chemistry as a practice of emancipatory education, instead of the pedagogy of knowledge transfer, through a liberating practice.

"The United Nations Educational, Scientific and Cultural Organisation (UNESCO), in its document "The Four Pillars of Education", stresses the importance of teaching that is geared to everyday life". According to the organisation, "It is up to education to provide, in some way, the maps of a complex and constantly agitated world and, at the same time, the compass that allows us to navigate through it". In this way, people will need a broad knowledge of the sciences, without which it will be impossible to face the difficulties that the world imposes.

The Seven Knowledges Necessary for the Education of the Future also emphasises the importance of knowing "knowledge", teaching what knowledge is and then passing it on. This knowledge must be broad and not fragmented, as is the case in today's education system, where the sciences are isolated, each with its own particularities. Thus, the interrelationship with other sciences will become indispensable for teaching chemistry. The reintegration of all parts of knowledge will make it possible to teach the human condition, which should be the essential objective of all teaching.

The teaching of chemistry for citizenship should be contextualised, in which chemical knowledge is not the focus, but rather the preparation for the learning of a citizen who is aware of their role in the environment. This contrasts with the thinking of some science educators who, through the transfer of knowledge, simplistically seek to resolve theoretical and technical issues of derivatives, integrals and even nuclear magnetic resonance spectroscopy, without taking into account their purpose, complexity, variability, subjectivity and social relevance.

In an attempt to delimit the focus of the research, we considered the current educational situation in the country and around the world. This situation suggests that teachers should consider in their practice the formation of critical, active citizens who possess systemic knowledge, and to this end they suggest that teachers establish a relationship between scientific knowledge and the student's reality. The basic question of this investigation concerns the students' conceptions of contextualisation in chemistry teaching and their teaching practice.

1.1   NATURE OF THE RESEARCH OBJECT

One of the biggest challenges of teaching chemistry in primary and secondary schools is to build a bridge between school knowledge and the everyday world of the students. Often, the absence of this link is responsible for apathy and distance between students and teachers (Valadares, 2001). By restricting teaching to a strictly formal approach, the various possibilities for making chemistry more "tangible" are overlooked and the opportunity to associate it with technological advances that directly affect society is lost (Chassot, 1993).

1.2   Historical context of IF Baiano

The Federal Institute of Education, Science and Technology of Bahia - Guanambi Campus was created as the Antônio José Teixeira Federal Agrotechnical School (EAFAJT) in 1993. However, it began its activities in 1995 with a high school Agricultural Technician course. Subsequently, the subsequent Agriculture and Zootechny courses were created. In 2007, a technical course in IT was created in the form of education for young people and adults, and in 2008, a technical course in Agroindustry was created. In December 2008, with the reformulation of the professional education system proposed by the federal government, EAFAJT was renamed the Federal Institute of Education, Science and Technology of Bahia, Guanambi Campus. As a result, it began to offer higher education courses in Agronomy, Chemistry, Agroindustry Technology and Systems Analysis and Development. The institute currently has a total of 1,150 students on the various courses on offer.

1.3   Research Problem

Unlike in other times, contemporary changes unleash a noisy whirlwind around us and affect the entire world population at the same time. Any significant event in the contemporary world is no longer distant. What used to be someone else's problem has become everyone's problem, because few major events fail to affect us in some way.

These processes have an impact on the future projects of trainers and trainees, due to new social expectations and demands in relation to education. There is increasing pressure for education with more and better results, as there is growing confidence in the value of quality education as an instrument for social development. Life in society is no longer the prerogative of geography and history lessons. Each subject in the curriculum, individually and collectively, must be attuned to the same goal: to favour the process of developing the individual as a human being.

At this point, a parenthesis opens up to reflect on the role of chemistry as a subject in the basic education curriculum. It is common to attribute to this discipline responsibility for the development of materials and technological advances that contribute so much to improving the quality of life. However, that same quality of life is affected by the shape of the current economic systems, structural changes in social organisation and the production of consumer goods in a disjointed and unbalanced way. And so, once again, the spotlight is turned on chemistry, but this time with censure, as if the science were responsible for the harmful effects caused to the environment. This issue cannot be left out of school. Chemistry urgently needs to bring this issue into the classroom, allowing for a broad and meaningful approach to the subject. The hope is that, on completing basic education, students will be critical citizens capable of reflecting on their role in society.

However, it has been observed that chemistry teaching has been developed in such a way as to value the memorisation of names and formulas without any relation to the students' daily lives. In this context, the central question is: has the lack of connection between curricular content and students' everyday lives contributed to their civic education?

1.4   Justification

Citizenship education is also an education of the human conscience in its ethical and moral values and the construction of a critical view of the world. Values that need to be founded on the principle of respect for life and the principle of equality, so that fundamental human rights are guaranteed, while at the same time there is a duty to commit to the new society.

Life itself is already a fantastic chemical process, in which the transformations of substances allow us to walk, think and feel. The various biological sensations, such as pain, cramps and appetite, and the various psychological reactions, such as fear, joy and happiness, are associated with the substances present in our bodies. Our body is a veritable laboratory of chemical transformations. Studying chemistry is important for every citizen to be concerned not only with facts of interest to themselves, but also those of interest to society. Scientific information helps us to understand events and to position ourselves as informed and aware citizens.

The more people realise that chemistry is a science that helps us to better understand the world we live in and that, if chemical knowledge is used wisely and ethically, there will undoubtedly be an improvement in the living conditions of all citizens. It is thanks to the theoretical evolution of chemistry that a series of substances that are commonplace in today's world have been achieved. Thus, from the chips used in computers to the metal or plastic prostheses increasingly used in medicine, from styrofoam to the fuels used in transport, everything is based on chemical knowledge, which is the achievement of research and study by countless generations of scientists.

Studying chemistry not only allows us to understand current phenomena. It helps us to understand the complex social environment in which we live. Science advances according to the needs generated by society, such as malnutrition, lack of energy, pollution, etc. In turn, technological improvement contributes to the development of the sciences and is also associated with scientific development. Chemistry has ensured that human beings live longer and more comfortable lives. Its development has made it possible to find solutions to environmental problems, treat previously incurable diseases, increase agricultural production, build sturdier buildings and produce materials that make it possible to manufacture new equipment.

Science, technology and society have been working towards solutions to major problems, but they have also had disastrous consequences for human life on the planet. Every day we read news stories showing the paradox of scientific and technological development, which brings both benefits to society and risks to human survival itself. According to Albert Einstein, quoted by Santos (2005. p.24) "Science has no meaning except when it serves the interests of humanity". In order to change this situation, all of us citizens should seek to develop actions in our community so that the applications of science and technology in society can protect the lives of future generations and provide the conditions for everyone to have access to its benefits.

From this perspective, chemistry teaching should contribute to the concept of science as a human activity under construction, with a view to understanding the student's socio-economic reality. In order to achieve this goal, it is necessary to reorganise curricular content in order to broadly construct the concept of chemistry and its social role. Thus, this project aims to investigate whether chemistry lessons contribute to students' civic education, according to the students' concept of integrated secondary education at the IF Baiano Guanambi Campus.

1.5 Objective:

1.5.1 General Objective:

To analyse the contribution of chemistry teaching to the civic education of 1st, 2nd and 3rd year Integrated High School students at the Federal Institute of Education, Science and Technology of Bahia - Guanambi Campus - BA.

1.5.2 Specific Objectives:

- To diagnose whether chemistry teaching at the Federal Institute of Education, Science and Technology of Bahia - Guanambi Campus - BA, has been favouring the interrelationship between the subject and the student's life;
- To see if the teaching of chemistry at the Federal Institute of Education, Science and Technology of Bahia - Guanambi Campus - BA, enables students to develop a critical view of the world;
- To find out, from the students' point of view, whether the teacher uses practical lessons to improve understanding of the theory.

CHAPTER 2

2.1 THEORETICAL BACKGROUND

2.2 History

Historians commonly consider that chemistry only became a scientific discipline in the 18th century, in a process that culminated with the work of Lavoisier. However, practical knowledge related to chemistry already existed. The Scientific Revolution, which took place in Europe between the 16th and 18th centuries, was a historical phenomenon of extraordinary dimensions.

In Brazil, the practice of science as an organised and regular activity only emerged belatedly and the road to institutionalising science in the country was long and arduous. The Jesuits established a network of educational institutions along the Brazilian coast. Several Jesuit colleges awarded licentiate and master's degrees, but without legal validity, as the University of Coimbra had not allowed Brazilian institutions to become universities.

In Salvador, BA, between 1670 and 1681, the Jesuits made several attempts to bring a university to the city, sending letters to the king of Portugal to get him to approve the idea, but they were unsuccessful.

The first official chemistry course was offered by the Chemistry Institute in Rio de Janeiro in 1918. In 1920, the Agricultural Industrial Chemistry course was created in association with the Higher School of Agriculture and Veterinary Medicine, and in 1933 it formed the National School of Chemistry in Rio de Janeiro. Today there are a large number of chemistry courses offered by higher education institutions in Brazil.

Current chemistry teaching is marked by the historical, economic and social influences that have taken place in Brazil. For many years, science was considered "neutral", *and* its importance only for practical uses. One hypothesis for this was raised in the work of Lops (2007, p. 82), who points out that historically the sciences were associated with the idea of "doing" and not "thinking" and had the role of preparing people for work, while literate knowledge had the role of preparing the spirit, i.e. science was not important for humanistic training, it was only technically useful. This conception, according to the author, may have given the sciences a descriptive character (descriptivism) and so teaching science consisted only of teaching facts and principles of practical use.

For a long time, the positivist view of science also permeated, in the case of chemistry, whereby students studying chemical phenomena had to be convinced by observation and experimentation that the laws and theories were correct. In addition, Lopes (2007, p. 96) highlights remnants of descriptivism, since "teaching chemistry meant describing the characteristics and properties of the most commonly used substances", reducing teaching to a simple application of content.

We therefore realise that the conceptions of science teaching that are still present in schools are the result of different social conditions and the different education reforms that have taken place at different times. Between 1960 and 1980, environmental problems, increasing pollution, the energy crisis and social movements led to profound changes in the proposals for science subjects at all levels of education. Connections between science and society began to be recognised, broadening the vision of science teaching. We began to believe in and defend the importance of linking science teaching to politics, economics and culture.

As a result of this movement, we began to realise the need to relate science teaching to everyday life, to students' experiences, bringing the importance of understanding the problems faced by society, as well as its relations with the world interconnected by communication systems and technologies. Allied to this, a vision of science was being created that went beyond the concept of discipline alone. In the current National Curriculum Parameters, many of the themes linked to science teaching are now considered "transversal themes", such as environmental education, health and sex education. However, as Krasilchik (2000, p. 89) states, "the school tradition still determines that the responsibility for teaching science falls basically to the scientific disciplines.

In this context, despite the various teaching reforms in Brazil, through which chemistry teaching has also been reformulated and rebuilt, what we still find today in most of our schools are traditional teaching practices characterised as theoretical, content-based and fragmented, distancing them from their experimental and investigative nature, as well as from real, everyday facts. For many students, chemistry lessons contribute little to their education and their chemical understanding of the world, as they consist of the practice of reproducing a few rules that are memorised in order to be described in tests.

As a result, they do not realise the important contribution of chemical knowledge to society and do not understand it in their daily lives. In the words of Maldaner:

> the current teaching of chemistry in schools does not provide students with the learning that enables them to understand the chemical processes themselves and to construct chemical knowledge, relating it to the cultural and natural environment, to environmental, social, economic, scientific and technological issues (MALDANER, 2006, p.61-2).

It is important to emphasise that in the area of chemistry teaching, much progress has been made in the last 20 years in proposals to overcome traditional teaching. This research is linked to groups from different educational institutions and is seen as an alternative for more meaningful and contextualised chemistry teaching.

2.2 The National Curriculum Guidelines for Chemistry Teaching

The National Curriculum Guidelines for Teaching advocate the need to contextualise teaching content with the reality experienced by students, in order to give them meaning and thus contribute to learning (Brazil, 1999).

> Chemistry can be an instrument of human education, broadening cultural horizons and autonomy in the exercise of citizenship, if chemical knowledge is promoted as one of the means of interpreting the world and intervening in reality, if it is presented as a science, with its own concepts, methods and languages, and as a historical construction, related to technological development and the many aspects of life in society. (PCN, 1997, p.187)

From this context, high school students should have an understanding of chemical processes in close relation to their technological, environmental and social applications, so that they can make value judgements and decisions in a responsible and critical manner. For this to happen, the learning of content must be associated with competences related to the pillars of education, namely: doing, knowing, knowing how to be and knowing how to be in society. For these objectives to be achieved, the selection and organisation of the content to be taught is a very important aspect.

Teaching and learning strategies should enable students to participate actively in lessons, through activities that challenge them to think, to analyse situations using chemical knowledge, to propose explanations and solutions and to criticise decisions constructively. In short, they should favour the formation of individuals who know how to interact more consciously and ethically with the world in which they live, in other words, with nature and society. The contents to be developed should be thought of by the teacher as structuring elements of the pedagogical action, i.e. it is not enough to explain the specific chemistry topics to be taught, but also to point out the learning expectations for each of them, their interrelationships and their applications for a better understanding of different contexts.

In the area of contextualisation and action, chemistry teaching should be carried out in such a way that students can understand science and technology as integral parts of contemporary human culture, recognise and evaluate their development and their relationship with the sciences, their role in human life, their presence in the everyday world and their impact on social life; recognise and evaluate the ethical nature of scientific and technological knowledge and use this knowledge in the exercise of citizenship.

The educator must be a mediator of this knowledge and not just a transmitter, who constantly seeks to improve himself, in other words, a teacher-researcher who is in the midst of continuous training to facilitate his pedagogical practice. The teacher must not forget that the social reality in which the students live must be taken into account and their previous knowledge must serve as a basis for the proposed menu. Before starting work, the teacher must apply an investigative method to gather data and also get to know the students and their social life.

2.3 The teaching-learning process

According to Law No. 9394/96 (Lei De Diretrizes E Bases, 1996) in Art. 1, Education encompasses formative processes that develop in family life, in coexistence, at work, in educational and research institutions, in social movements and civil society organisations and in cultural manifestations. School education must exercise democracy and citizenship, as a social right, through the appropriation and production of knowledge. School is influenced by its environment, it is not neutral, it is the result of the actions, values and principles of the historical reality that interferes in its procedures. "School prepares, instrumentalises and provides conditions for the construction of citizenship, for the formation of a critical citizen who is the subject of their own history." (LIBÂNEO, 1993, p.33).

In fact, education is precisely a process of transformation and natural inclinations of the subject, in the face of what society already constitutes as properly human, that is, as culture imposes on education the understanding that man is not a juxtaposition of biological, psychological characteristics.

> Teaching is, first and foremost, handcrafting knowledge so that it can be taught, practised and assessed within the framework of a class, a year, a timetable, a system of communication and work. [...] It's important to note that in order to be taught, acquired and assessed, knowledge undergoes transformations: segmentation, cuts, progression based on pre-constructed materials (manuals, brochures, worksheets). In addition, it must be part of a viable didactic contact, which establishes the status of knowledge, ignorance, error, effort, attention, originality, questions and answers.
>
> (PERRENOULD apud KULLOK, M. G. B, 2002, p.10)

The process of teaching implies a new way of conceiving the classroom, which should only be a place for transmission, but above all a space for constructing knowledge. For this to happen, teachers need to review their way of teaching and conceiving of teaching. However, when we look at the process of learning, it is seen as seeking information, reviewing one's own experience, acquiring skills, adapting to change, discovering meaning in beings, facts, phenomena and events, modifying attitudes and behaviours.

All these actions point to the student as the main agent responsible for learning. With this, the teacher is concerned with what the student needs to learn in order to become a citizen: how the student will learn best, what techniques will favour the student's learning: how the assessment will be carried out in order to encourage the student to learn. Learning is the process by which the subject actively appropriates existing content. Learning is much more meaningful as the new content is incorporated into a student's knowledge structures and acquires meaning for them from the relationship with their prior knowledge.

When the school content to be learnt cannot be linked to something already known, mechanical learning occurs, i.e. when new information is learnt without interacting with relevant concepts in the cognitive structure. Thus, the person memorises formulas and laws, but forgets them after the assessment. For meaningful learning to take place, two conditions are necessary. Firstly, the student needs to be willing to learn: if the student wants to memorise the content arbitrarily and literally, then learning will be mechanical.

Secondly, the school content to be learnt has to be potentially meaningful, that is, it has to be logically and psychologically meaningful, logical meaning depends on the nature of the content, and psychological meaning is an experience that each individual has. Each learner filters out the content that is or is not meaningful to them. In the teaching-learning process, the student is the subject and the builder of the process. All learning needs to be based on a good relationship between the elements that take part in the process, i.e. the student, teacher, classmates: dialogue, collaboration, participation, joint or group work and games, mutual respect, etc.

The way in which these three elements (teacher, student, programme) interact will reveal, for example, the teacher's conception of learning and the teaching-learning process: their role in it, the role of the student, their view of the world and contemporary society: It will reveal ways of integrating theory and practice, science and everyday reality outside the school structure, it will indicate the political and educational guidelines of both the MEC and the concrete institution where the classroom takes place: in other words, the classroom is a small world where, in the actions and interactions of the teachers-students programme on a daily basis, the education of our students-educators takes place.

However, to this day, the emphasis has been on the teaching process and, with this, the concern is with the organisation of the curriculum in the form of watertight and isolated subjects: the methodology favours oral transmission without a search for the learning process, but focused almost exclusively on complying with the established programme; and the teacher is seen only as the one who "masters the content" regardless of how he or she transmits it. In this case, teachers often ask themselves: what do I think is important to teach? How am I going to teach? How do I like or prefer to teach? How I find it easiest to teach.

In the learning process, the questions teachers ask themselves are also different: what do students need to learn in order to become citizens? How will the student learn best? What techniques will favour the student's learning? How will assessment be carried out to encourage learning? In this approach, the teaching and learning processes are distinct. Emphasising one or the other will mean that the results of integrating or correlating the two processes will be completely different. To date, the emphasis has been on the teaching process. However, it is necessary to change the process of learning, various habits and attitudes in order to promote scientific knowledge.

> Thus, according to Moraes (2008): The production of scientific knowledge, in general, occurs by questioning and reconstructively expanding existing knowledge and theories already accepted by a community of specialists in a field. In the same way, at school, learning takes place through the reconstruction and complexification of the knowledge that the student already brings to the school context, a process that begins with questioning and culminates in expanded understandings of the issues questioned. (Moraes, 2008, pg. 3)

The methodology of learning requires knowledge of the cognitive area, i.e. understanding how human beings think, reflect, analyse, compare, criticise, justify, argue, interfere, conclude, generalise, search for and process information, produce knowledge, discover, research, create, invent and imagine. It also requires knowledge of human skills, i.e. learning to communicate with colleagues and teachers, working in teams, participating in groups, writing and presenting work. There is no doubt that the different methodological proposals used by educators also have an impact on the teaching-learning process.

This objective can be achieved more effectively if the teacher has the competence to realise his or her role in this process, and starts from the understanding that, in all fields where they are intended to be effective, it is necessary to create in the seeker the idea that he or she is a capable being: to motivate the student, to "raise his or her esteem". This is the first step towards transforming the student into a dynamic, conscious and participative person. In other words, understanding education as a process in which competence is realised, both for the teacher and the student.

In the etymological sense, the term competence comes from the Latin word competência, which means the quality of someone who is capable of appreciating and resolving a certain matter, doing a certain thing; capacity, ability, aptitude, suitability (FERREIRA, 1999).

> Competence is not a state, but a process. If competence is a way of acting, how does it work? A competent operator is one who is able to mobilise, to effectively apply the different functions of a system in which resources as diverse as reasoning operations, knowledge, memory activation, evaluations, relational capacities or behavioural schemes are involved. To a large extent, this alchemy remains an unknown quantity [...] Competence does not lie in the resources (knowledge, skills...) to be mobilised, but in the very mobilisation of these resources. Competence belongs to the order of knowing how to mobilise. For competences to exist, a repertoire of resources - knowledge, cognitive skills, relational skills - must be in play... (LE BOTERF apud PERRENOUD, 2001, p.13 and 21).

A competent teacher is one who has the ability to mobilise and update knowledge, i.e. it would be pointless for a teacher to have knowledge/know-how/skills in a particular subject if they didn't know how to apply, plan, develop and evaluate the various educational situations; if this were the case, there would be no competences, only skills (PERRENOUD, 2001). Skills are associated with knowing how to do: physical or mental action, which indicates acquired capacity. So identifying variables, understanding phenomena, relating information, analysing, synthesising, judging, correlating and manipulating are all examples of skills.

2.4 Education and Citizenship

Teaching is a systematic form of knowledge transmission used by humans to instruct and educate their fellow human beings, usually in places known as schools.

The teaching of chemistry and training for citizenship are linked within the framework of the law as the aim of basic education, as well as the technological influence of chemistry on modern society.

The changes that have been taking place in the world, mainly in relation to technology and the progress of science, have changed people's daily lives. Advances in communications, industry, transport and cybernetics have shortened the distances between people, changing concepts of space, time and learning. This has changed the school perspective on the construction of knowledge in this new context.

The new educational paradigm is to prepare individuals who can learn to think and act in order to interact with the world (citizens), and not just memorise content. With this in mind, the Ministry of Education has taken a series of measures to give Brazilian education the qualities needed for this change, the most important of which is the National Education Guidelines and Bases Law (LDB) - Law No. 9.394/1996.

Learning chemistry doesn't mean memorising the periodic table, much less making 'dots and stripes' to represent chemical bonds. Learning must be closely associated with the ability to relate to everyday life, the subject of chemistry must be contextualised with themes that relate to the everyday activities of the diversity present in the classroom, and the uniqueness of each student must be preserved. The chemistry teacher must bring dialogue to the class, and to do this it is necessary to bring his or her own experience into this sphere. Individual experiences foster the development of meaningful learning.

Chemistry must offer the possibility of improvements, the possibility of aiming for changes in the way we see and think about the world, and allow for more questions than end points. It is necessary to question the quality of the simplest day-to-day activities, and it is through questioning that changes are brought about, which is why it is essential to rethink the way content is organised and the way classes are conducted.

Chemistry teaching, as in other exact sciences, still generates a feeling of discomfort among students because of the learning difficulties that exist in the learning process. Chemistry is often taught in a traditional way, in a decontextualised and non-interdisciplinary way, generating a great lack of interest in the subject among students, as well as difficulties in learning and relating the content studied to everyday life, even though chemistry is present in reality.

Contrary to the traditional teaching model, it is argued that learning chemistry should enable students to understand the chemical transformations that occur in the physical world in a comprehensive and integrated way, so that they can judge it on theoretical and practical grounds (NUNES; ADORNI, 2010). However, teachers are not always prepared to act in an interdisciplinary way, relating the content to the students' reality, making it necessary to look at chemistry teaching in terms of citizenship.

Teaching chemistry to train citizens, in short, means teaching chemistry content with the primary aim of developing the student's ability to participate critically in society's issues, in other words, "the ability to make decisions based on information and weighing up the various consequences of such a position" (SANTOS and SCHETZLER, 1996, p. 29). Chemistry is a science that is constantly present in our society, in consumed products, in medicines and medical treatments, in food, in fuels, in energy generation, in advertising, in technology, in the environment, in the consequences for the economy and so on. Citizens are therefore required to have a minimum knowledge of chemistry in order to participate in today's technological society.

2.5 Teaching for citizenship

Citizenship has been and is under permanent construction; it is a benchmark for the conquest of humanity, through those who always fight for more rights, greater freedom, better individual and collective guarantees, and who do not conform in the face of arrogant domination, whether by the state itself or by other institutions or people who do not give up on privileges, oppression and injustices against an unassisted majority who cannot make themselves heard, precisely because they are denied full citizenship whose conquest, however late, will not be hindered.

To be a citizen is to realise that you have rights. Rights to life, liberty, property, equality - in short, civil, political and social rights. But this is just one side of the coin. Citizenship also presupposes duties. Citizens must be aware of their responsibilities as an integral part of a large and complex organism that is the community, the nation, the state, to whose proper functioning everyone must contribute. This is the only way to reach the final, collective goal: justice in its broadest sense, i.e. the common good.

The school today, at least from a theoretical perspective, is strongly committed to quality teaching and the idea of building citizenship. The school content taught to students is seen as part of the tools needed for everyone to understand the reality around them and acquire the necessary conditions to discuss, debate, give their opinion and even intervene in the social issues that mark each historical moment. According to the National Curriculum Parameters (PCN):

> The quality education that society is currently demanding is expressed here as the possibility for the education system to propose an educational practice that is appropriate to the social, political, economic and cultural needs of the Brazilian reality, that considers the interests and motivations of the students and guarantees the learning that is essential for the formation of autonomous, critical and participative citizens, capable of acting with competence, dignity and responsibility in the society in which they live. (PCN, 1997, p.187)

Thus, a question that is currently the subject of much discussion among teachers is this: what does it mean to educate for citizenship? Or rather, how can we educate autonomous, critical and participative citizens in the classroom? These questions arise for a very simple reason: no teacher disagrees with these ideals. The problem is that, as well as being an extremely broad idea, according to Demo (1993), not every teacher or school is sufficiently prepared to create the practical conditions needed to form citizens capable of acting competently and consciously in society.

Chemistry in secondary education is treated in this way by the authors of the book Educação em Química: Compromisso com a cidadania (Chemistry Education: Commitment to Citizenship), where a citizen does not need to have such specific knowledge of chemistry in order to live better in society. It may be interesting and advisable for citizens to have this knowledge for their cultural enrichment, but chemistry in secondary education cannot be taught for its own sake, otherwise we will be running away from the greater purpose of basic education, which is to ensure that individuals have the training that will enable them to participate as citizens in life in society. Once again, the importance of teaching chemistry for the formation of citizens is emphasised.

For Piaget (1973), an individual's knowledge is not something born, but is built from interaction with the environment, so the greater the number of complex interactions, the more intelligent the individual will be. Therefore, knowledge cannot come about by merely copying reality, but rather through different everyday situations that are important in the student's education.

Thus, chemistry education for citizenship aims to enable students to make decisions when solving problems that involve their lives and society, and this requires knowledge that goes beyond the field of science and that values a political perspective and evaluative attitudes.From this point of view, teaching cannot be based on classes that only require the memorisation of names and formulas or repeated training in solving standard problems, making it uninteresting, but rather through well-prepared and interspersed classes, well contextualised, with experiments that awaken the practical sense of things, motivating the student to learn.

2.6 Teaching Chemistry and Society

In the past, the context of the expansion and massification of the education system favoured the tradition of secondary education being preparatory for entry into higher education and the standardisation of textbooks. This was done in order to comply with university entrance exam programmes, not only in terms of content, but also in terms of the way they were approached.

Secondary education is the level of schooling where its social function is most unclear. The dichotomy between the terminal character and the continuity of secondary education (vocational or preparatory function for higher education) has led to a division between the educational establishments themselves, reproducing the social division within the school.

It was in this context that the proposal for basic education emerged, with the aim of shaping citizenship. In it, secondary education has the role of completing the basic education common to all Brazilian citizens, ensuring the minimum equal education required for effective participation in society. In theory, this idea seeks to overcome the aforementioned dichotomy. In practice, however, high school continues to be seen by many students and teachers exclusively as a preparatory course for university entrance, which contributes to de-characterising this level of education.

This reductionism leads schools to lose their function of forming critical citizens who are aware of their country's social reality and willing to transform it. The pressure of the entrance exam on teaching curtails the teacher's pedagogical work, encouraging the memorisation of rules, the resolution of numerical exercises and the summary study of extensive syllabuses, to the detriment of precise conceptual understanding and an understanding of their relationship with the various fields of knowledge.

We must recognise that many Brazilian universities have modified their entrance exams, changing the syllabus, avoiding memorisation questions, favouring reflective questions and valuing conceptual understanding. These changes have not always been accompanied by schools, which, in the majority of cases, continue to follow an outdated script, overloaded with content and demanding levels of detail. Research into chemistry teaching has demonstrated the ineffectiveness of such programmes and methodologies. A new educational reform is gradually taking hold and the time is coming for secondary schools to assume their identity and advance their formative character.

Modern society requires a lot more knowledge and skills from our students. And it is at this point that knowledge of chemistry reveals its great importance, because we live in a technological society that demands from its citizens attitudes towards a development model that guarantees the existence of future generations. This implies understanding a minimum level of knowledge in order to understand the role of science, technology and their social interrelationships, and to develop attitudes and values.

The development of values in citizenship training means respect for aesthetic, political and ethical principles, encompassing the aesthetics of sensitivity, the politics of equality and the ethics of identity: organising principles of secondary education. To this end, the discussion of aspects related to science and technology, known as socio-scientific aspects, must be interwoven with the socio-cultural values that underpin national education.Preparing students for the conscious exercise of citizenship and for entry into higher education are not mutually exclusive objectives. It's a mistake to think that students will only have a chance at the entrance exam if their teacher trains them all the time to take the exams, and it's also a mistake to think that, when preparing for the entrance exam, dense programmes are needed with no room for an experimental and social approach to chemistry. On the contrary: when we provide students with a broader education, we are helping them to consolidate concepts and develop the logical reasoning required for entrance exams.

2.7 Contextualisation as an understanding of everyday life

Currently, in relation to teaching, the term everyday has been characterised as a study of ordinary situations linked to people's day-to-day lives. The function of teaching, from this perspective, is to relate knowledge linked to the student's daily life with scientific knowledge. The everyday approach can be characterised as teaching content related to everyday phenomena and facts with a view to learning concepts (DELIZOICOV et al., 2002, SANTOS and MORTIMER, 1999b).

Chassot (2001) points out that teaching that promotes the study of everyday life has become something of a fad and that it has the purpose of teaching scientific concepts pure and simple. For the author, there is a reductionism in this perspective of contextualisation.

This way of approaching everyday life in chemistry teaching is also presented in the PCN (BRASIL, 2002, p. 93):

> The content organisation proposal presented (...) takes into account two perspectives for teaching chemistry that are present in the PCNEM: the one that considers the individual experience of the students - their school knowledge, personal histories, cultural traditions, relationship with everyday facts and phenomena and information conveyed by the media; and the one that considers society in its interaction with the world, highlighting how scientific and technological knowledge has been interfering in production, culture and the environment.

Thus, in view of this reductionist perspective, everyday life in this work will be characterised as the predominantly scientific study of everyday facts and phenomena. This approach is supported by the first three ideas of everyday life presented by Lutfi (1992).

A pedagogical practice based on the use of everyday facts to teach scientific content can characterise everyday life in a secondary role, i.e. everyday life is used as an example or illustration to teach chemical knowledge. Jiménez Lizo et al. (2002) point out that studies from this perspective use everyday phenomena in lessons as examples immersed in theoretical scientific knowledge in an attempt to make them more comprehensible. The authors emphasise that scientific literacy from the point of view of the study of everyday life has fallen into the reductionism of increasing the number of everyday examples in lessons.

A striking feature of the use of everyday aspects in chemistry teaching is the belief in their motivational potential, i.e. everyday situations, when exemplified, serve to motivate students to learn. Generally, these situations are introductory to the theoretical content and are intended to attract the student's attention, pique their curiosity, but are exclusively motivational, with the sole purpose of teaching content (CAJAS, 2001; LUTFI, 1992).

One justification for this reductionist view of the study of everyday life is also related to the instructional material that teachers use in their classes. According to Wartha (2005), most of the teaching materials used to support the teacher's lessons present a simplistic view of contextualisation, which is restricted to immediate everyday life. Thus, adopting the study of everyday phenomena and facts can lead to an analysis of situations experienced by students and teachers, which, due to various factors, are not problematised and consequently not analysed in a more systemic dimension as part of the physical and social world.

2.8 Teaching Chemistry in Guanambi

The municipality of Guanambi suffers from an extreme shortage of teachers with a degree in chemistry. The subject in the state network is taught by teachers from other areas and trainees. The reality is quite critical, as the lack of training leads to precarious teaching that is insufficient for students' education and citizen training.

Today, the region is home to the Federal Institute of Education, Science and Technology of Bahia, which has a large staff of teachers with a degree in chemistry (4 where; 3 with a master's degree and 1 specialist), and in the other areas, the majority with a master's degree, or studying for a master's degree and a doctorate.

Within this context, the formation of critical citizens is lost due to the difficulty of finding someone to teach the subject and applying coherence, pedagogical practice and the student's social reality.

CHAPTER 3

METHODOLOGY

For the development of this project, a qualitative-quantitative research methodology was chosen.

To this end, in the first stage, questionnaires containing 10 objective questions were administered to 5 (five) students, chosen at random, from each of the classes (1st, 2nd and 3rd year) of the integrated courses in Agriculture and Agroindustry. The results of the questions were expressed as a percentage. The second stage consisted of bibliographical research on citizenship and education through documents and forms of teaching aimed at citizen education.

An extremely important stage in a research project is the way in which the researcher looks at the data, and this concern was part of this work. According to André and Ludke (1986), data analysis involves, at first, processing the data in order to identify relevant trends and patterns. In the second stage, the trends and patterns that have been verified are re-evaluated, seeking relationships and inferences at a higher level of abstraction, with the aim of linking them to theorising.

3.1 Qualitative analyses

The need for qualitative analyses in this research is justified by the very complex activity of the teaching job (educator), his relationship with the student, with the knowledge to be taught, as well as the form he adopts to teach.

Bogdan and Biklen (1994) describe five characteristics of qualitative research which, in a way, guided the sequence of work carried out in this study. These are: the natural environment as the source of the data; care in describing the data in words or images; the systematic study of the process and not just the results; analysis and abstractions as the data is grouped. Thus, according to the authors LUDKE 1982 and ANDRÉ, 1986, qualitative research involves obtaining descriptive data for analysis through the researcher's interaction with the context of study, emphasising the process rather than the process itself.

the results and is concerned with portraying the perspectives of those surveyed.

3.1.1 Case Study

It consists of using one or more quantitative methods to gather information and does not follow a rigid line of enquiry. It is characterised by describing an event or case. The case generally consists of an in-depth study of an individual unit, such as: a person, a group of people, an institution, a cultural event, etc. As for the type of cases studied, they can be exploratory, descriptive or explanatory. The questionnaire, according to Rodríguez (1999), cannot be said to be one of the most representative techniques in qualitative research, as its use is more associated with quantitative research techniques. However, as a data collection technique, the questionnaire can provide an important service to qualitative research. This technique is based on the creation of a pre-prepared and standardised form.

3.2 Quantitative analyses

Quantitative analysis is better suited to ascertaining the explicit and conscious opinions and attitudes of those interviewed, as it uses structured instruments (questionnaires). It must be representative of a given universe so that its data can be generalised and projected onto that universe. Its aim is to measure and allow hypotheses to be tested, since the results are concrete and less prone to misinterpretation. In many cases, indices are created that can be compared over time, making it possible to trace a history of information.

It is appropriate when there is the possibility of quantifiable measurements of variables and inferences based on numerical samples, or when looking for numerical patterns related to everyday concepts. In this sense, in order to methodologically subsidise the case study, we sought the "qualitative-quantitative" methodological approach (Trivinos, 1987), in which both the quantitative dimensions, referring to the treatment of statistical data, and the qualitative dimensions, responsible for the interpretative quality of the information, are integral parts of the foundations responsible for interpreting the phenomenon being researched.

CHAPTER 4

RESULTS AND DISCUSSION

At the end of the discussion and reflection stages and the analysis of the materials, this research shows an Inefficiency in the teaching of Chemistry at IF Baiano - Campus Guanambi in terms of its conception aimed at the critical and citizen training of its students, who initially characterise the teaching of contextualised Chemistry as a simple exemplification and description of everyday facts or situations without the intention of teaching and understanding ideas of contextualisation such as addressing social issues, with a view to developing attitudes and values and transforming social reality.The first question investigated the student's interest in chemistry (Figure 1).

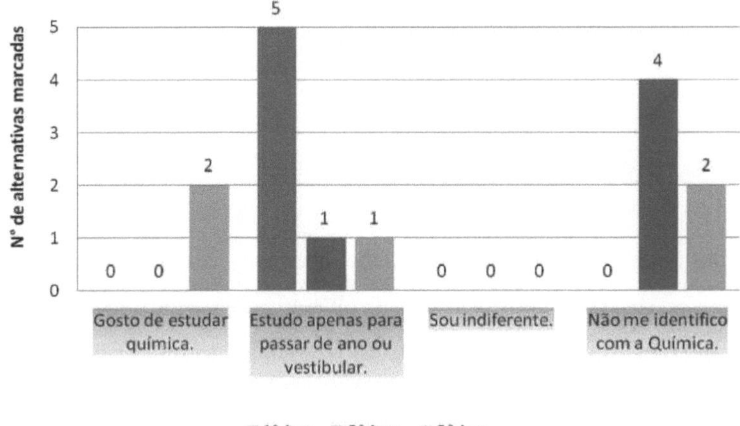

Figura 1: What is your interest in chemistry?

The responses were worrying, since 100 per cent of first-year students said that they only study to pass the entrance exam. This context is considered flawed, as it only deals with technical training and little application, since the "head full of knowledge" desired by students and even by the school itself is no guarantee that this knowledge will be applied in the student's daily life, making chemistry as a subject nothing more than concepts, mathematical calculations and chemical formulae without applicability.

Furthermore, we have to consider that the institute is a vocational school. The students' responses lead us to believe that even this objective is not being achieved. Even more worrying is the situation in 2nd year - where four students said they didn't identify with the subject, given that these students had already had one year of chemistry. The second question looked at the importance and applicability of chemistry teaching for the student (Figure 2).

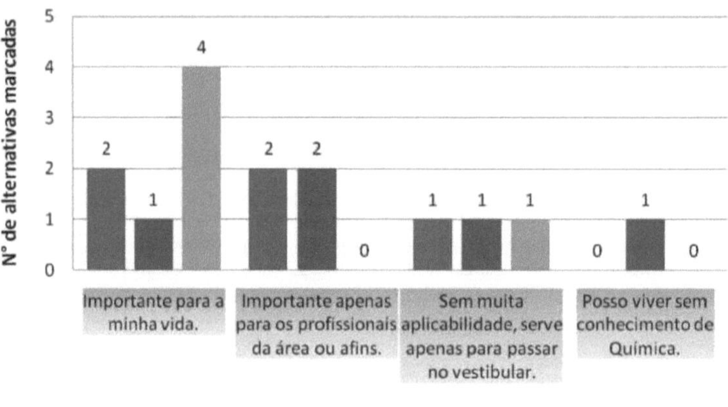

Figura 2:   What is the student's view of chemistry teaching?

Of the five third-year students, four replied that chemistry is important for their lives. At least on this point, teaching is fulfilling its objective. It is worth highlighting the differences between the responses of 1st and 2nd year students on the first item. A greater number of 1st year students consider the subject to be important for their lives, which shows that chemical knowledge, along the lines of education for citizenship, is still lacking and begs the question: did the 1st year student acquire this ability in their grade or did they already bring it from primary school? This is because the student has contact with chemistry in the ninth year of primary school. Following on from the analysis of the forms, question number 3 investigated the methodology used by the teacher in the classroom in relation to the subject of Chemistry (Figure 3).

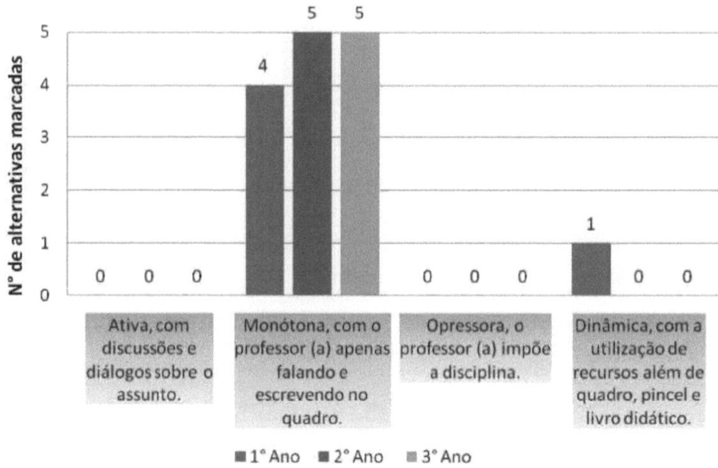

Figura 3:   What methodology does the chemistry teacher use?

It showed what scholars of the teaching of the subject have most warned about: the teacher is still the centre of knowledge, he/she is simply a purveyor of knowledge and does not give the student the opportunity to participate. Banking education, which Paulo Freire fought so hard against, is still what prevails, in the students' view, in everyday life at the IF Baiano Guanambi Campus.The students' responses regarding chemistry teaching and everyday life are shown in figure 4.

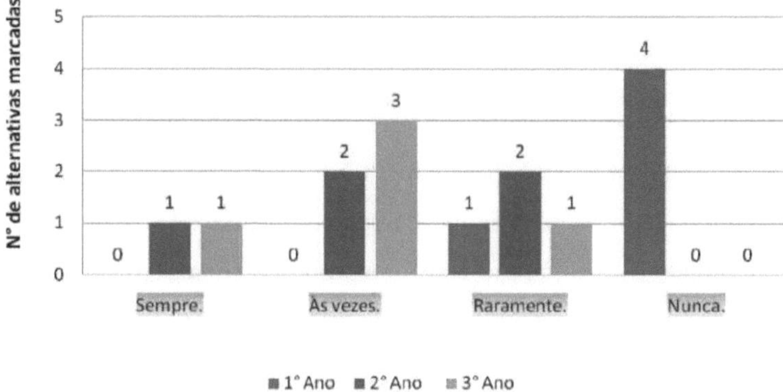

Figura 4:   Does the content taught by the teacher relate to everyday life?

For the minority of 2nd and 3rd year students, this relationship always happens.

29

The majority of students in these classes ticked the options sometimes and rarely, which, from the point of view of teaching quality, do not differ significantly. In the 1st year, the answer "never" was significant. In this class, the contents are basically atomic models, chemical bonds, the periodic table and inorganic compounds. This is fairly abstract content, but even so the teacher must endeavour to relate it to the students' everyday lives, otherwise chemistry teaching will be doomed to failure.One point that is always discussed by theorists and scholars of chemistry teaching is experimentation. All of them unanimously recognise practice as an instrument that facilitates the understanding of theory. When asked about the use of laboratory practice by the teacher, most of the students in the three classes said that they had never had laboratory lessons (Figure 5).

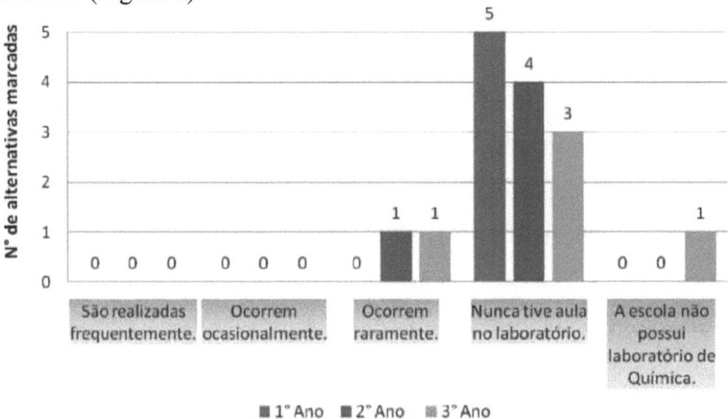

Figura 5: Do practical classes take place in the laboratory?

It is also noticeable that little use is made of the PCNs, which encourage practice aimed at citizen education. Chemistry teaching at IF Baiano - Campus Guanambi has prioritised theory, which makes it difficult to learn the proposed content; theoretical, without correlation with practice, making it difficult for students to learn the proposed content.

Question 6 assessed whether, in the students' view, the knowledge acquired in the classroom makes it possible to evaluate the social implications of the technological applications of chemistry in everyday life (Figure 6).

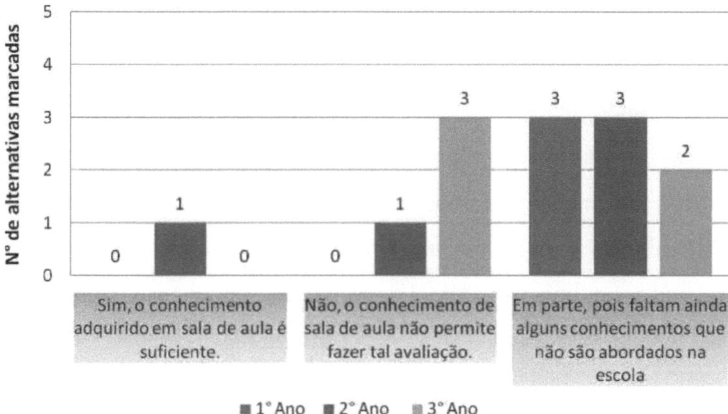

Figura 6: Does the knowledge acquired in the classroom allow you to evaluate the social implications of chemistry in your day-to-day life?

The majority of third year students said no. It's worth noting that these students had already taken the subject in two previous years. Is it possible to see a slight improvement in the 1st and 2nd year classes who said partly? From what has been observed in the answers so far, we could certainly refute this question. The answer may also indicate, when the student says that some knowledge that is not covered at school is still lacking, that the student is aware of this knowledge or has acquired it elsewhere. However, according to Graph 2, the students emphasised that there is a need to acquire knowledge of chemistry, as it is considered important in their lives. This is an interesting fact, as it reduces the view of science as something unattainable, leaving it up to the student as a citizen to accept knowledge or not, not transferring their decision-making capacity to others.

Interdisciplinarity is constantly being discussed in education. It is no longer possible to conceive of an education where each subject is isolated from the others. Chemistry, for example, is the basis of everything. We are made up of atoms. The agglomeration of these atoms make up substances, the tissues studied in biology, or the agrochemicals used in agriculture.

The social implications of the use of these substances are excellent approaches for teaching any subject. However, what still prevails in Brazilian education, and what can also be observed at the IF Baiano Campus Guanambi, is content that is disconnected from each other. In other words, the subjects are like drawers in a cupboard where to open one you have to close the other. The data shown in figure 7 allows us to assume that the majority of students see the interrelationship between chemistry and the other subjects in their curricula. One relevant fact is that the 2nd year class had a different understanding from the 1st year class, which was somewhat unexpected, given that the class has been at school longer than the 1st year.

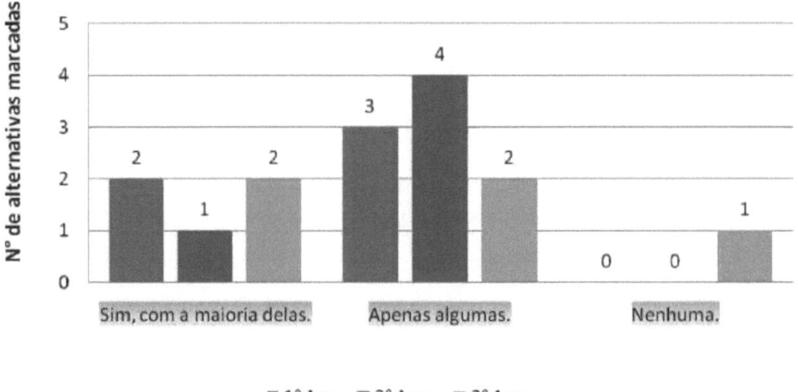

Figura 7: Is it possible to relate chemistry subjects to other subjects?

With regard to participatory behaviour, the students' responses (Figure 8) show that the teacher has encouraged participation by enabling open dialogue between the students and also uses the students' previous knowledge to develop the content.

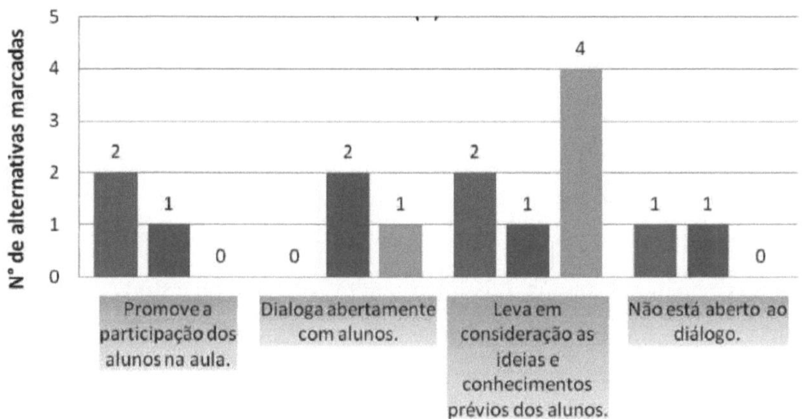

Figura 8: Does the teacher promote student participation or not?

Regarding the approach to environmental issues in the classroom, it can be seen, in the students' view, that the teacher rarely does this (Figure 9). The environment and science should be intrinsically linked. Through chemistry it is possible to understand the life cycle of products, the consequences of their disposal and the contributions that science can make to minimising environmental impacts. There is a wealth of approaches to teaching chemistry. The hardest part is raising awareness and, above all, preparing teachers for a radical change in the way they teach.

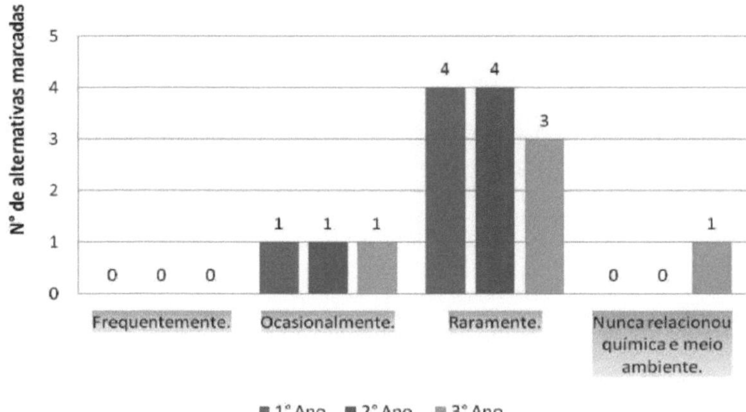

Figura 9: Has the chemistry teacher addressed environmental issues in the classroom?

Finally, the last question sought to find out how the examples given by the teacher in class were interpreted by the students (Figure 10). A certain discrepancy was observed between the answers given in the previous questions, when the student stated that the teacher rarely correlates chemistry to their daily lives. In this case, the most frequent response was that the examples given help to relate chemistry to everyday life.

It can be said that the way chemistry has been taught is not just the prerogative of the IF Baiano Guanambi Campus. It is important to comment on a study carried out by Brito (2008) with secondary school students, whose data collection instrument was used in this research. The results are very similar, which hides an even more serious problem that could be the subject of research at another time, despite being a constant theme in various studies: the training of chemistry teachers. You can't change the classroom without changing the root of the problem: teacher training, in any area, must be linked to the most urgent and necessary issues for reformulating teaching.

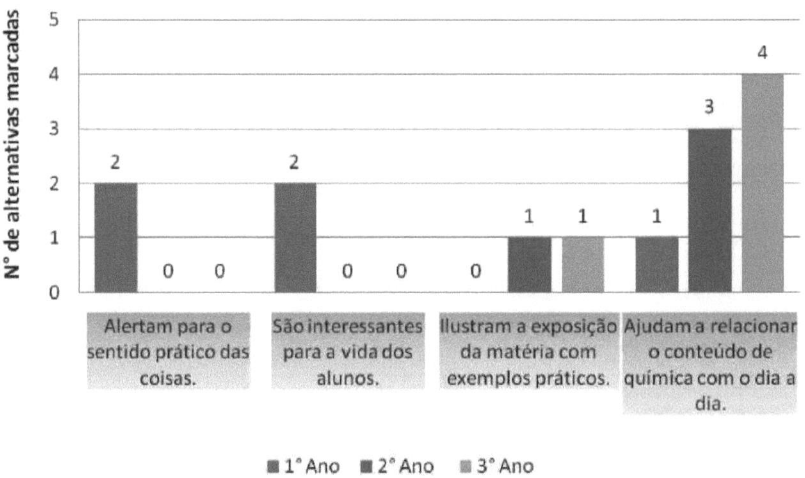

Figura 10:   Regarding the examples the teacher uses in class.

The contents and learning strategies should value the relationships between individuals, work, political relations and the subjectivity of each individual, and were used for the four premises pointed out by the United Nations Educational and Scientific Organisation (UNESCO) as guidelines for education in today's society.

The first of these, "learning to know", prioritises the fact of knowing, learning and understanding the phenomena of the environment. Those who have a considerable amount of knowledge can understand the world better, observing facts from different angles of knowledge, in a critical and autonomous way. The second premise is "learning to do", which is about the student obtaining skills from the practical application of theory, valuing the uses of science and technology in the social context.

"Learning to live" is the third premise. Man is a social being, even when it comes to science, man develops knowledge of others and this is a way of showing that they have common goals, enriching the scientific and technological world. Finally, "learning to be", which is about preparing the individual to deal with the different circumstances of life, through critical thinking, value judgement, discernment and feeling, as the possessor of their own destiny. These premises aim to put an end to the purely quantitative view and the accumulation of knowledge as an evaluative tool in the teaching-learning process.

In this sense, it would be more interesting for students at IF Baiano - Campus Guanambi to acquire knowledge that is closely related to their daily lives, such as: knowing how to handle chemical substances, interpreting information provided by the media, understanding the chemical language of food and medicines, having a critical opinion on social and environmental problems involving chemistry, among others.

CHAPTER 5

FINAL CONSIDERATIONS

In our opinion, contextualisation in chemistry teaching, which favours the study of social contexts with political, economic and environmental aspects, based on knowledge of science and technology, is the right way to develop teaching that will contribute to the formation of critical, active students who, whenever possible, will transform their unfavourable reality. To this end, contextualised teaching should not be limited to examples of facts, phenomena, processes, etc. Contextualisation as a scientific description of these aspects can guarantee more meaningful learning for students. Although this second view is unlikely to promote the development of attitudes and values or social transformation. However, this view can still be considered an advance on the first.

In this way, contextualised chemistry teaching must start from a broad context that can be studied on the basis of the knowledge developed. This study will certainly provide students with the tools to understand their physical and social environment with a view to intervening in it. To this end, it is necessary to ask the students to take a stand and act to change what they have studied, and for the teacher to have continuing training, in an effective search for scientific and world knowledge, so that he or she can add to the teaching of chemistry in training not just mere students, but citizens who are aware of their role as humans, with rights and duties.

Lutfi (1988, 1992) has a strong social orientation in his pedagogical proposal for teaching chemistry. The author believes in social theory in order to really understand what is happening. His proposals present arguments for discussion in this field, mainly because they make it possible to deal with social aspects, seeking to highlight possible political, economic and technological implications relating to scientific knowledge. In them, the focus is no longer essentially chemical, which, according to the author, allows for more meaningful learning for the student.

For Lutfi, everyday life assumes a degree of physical and social importance, which must be combined with other areas of knowledge, such as politics, economics, etc. In this vision of study, the individual is not alienated from his or her daily life, but needs new knowledge in order to understand the world around him or her and, if possible, transform it.

Based on Lutfi's thinking, with which we agree, this research points to the need for further studies that make more contributions to understanding the mechanisms of teacher action and the process of teachers recognising the effective epistemological, affective and cognitive barriers that control their choices and decisions in the educational process to train critical citizens.

Among the limiting factors of chemistry teaching for the exercise of citizenship, it is possible to highlight:

1°) Lack of understanding of the true meaning of teaching chemistry;

2°) Educators don't use the concept of science as a human activity under construction;

3°) The chemistry teachers at IF Baiano - Campus Guanambi, are failing to articulate an interconnection between the content and the student's prior knowledge;

4°) The study of chemistry has not contributed to students' development, promoting a critical view of the world around them;

5°) The organisation of curricular content is outdated for the development of students in the conscious exercise of citizenship;

6°) The precariousness of the broad concept of chemistry and its social role;

7°) Lack of public policies and their encouragement;

8°) Lack of a link between the content and the social context in which the student is inserted;

9°) Difficulty on the part of the educator himself in the conscious exercise of citizenship;

10°) A preoccupation with presenting a large amount of information in an attempt to cover all the content that textbooks traditionally cover, promoting mechanical learning of definitions and isolated laws, memorising formulas and equations.

Faced with this large number of factors that hinder the implementation of a democratic education geared towards the formation of citizenship, we can conclude that the teaching of chemistry for the formation of citizenship constitutes a new educational paradigm. And it is up to the teacher, as the active subject of the educational process, to adopt a political stance committed to transforming the subject of chemistry in the classroom into an instrument of construction and social justice, rather than the reproduction of existing economic and political models.

CHAPTER 6

BIBLIOGRAPHICAL REFERENCES

ANDRÉ. M. E. A.; LUDKE, M. Pesquisa em Educação: abordagens qualitativas. São Paulo: Ed. EPU, 1986. Basic Topics in Education and Teaching Collection.

BOGDAN R. BIKLEN S. Qualitative research in education: an introduction to theory and methods. Porto Editora. 1994 (Education Sciences Collection).

BRAZIL. Law of Guidelines and Bases of National Education, Law No. 9.394, of 20 December 1996.

BRAZIL. MEC. SEMTEC. (1997): Parâmetros Curriculares Nacionais para o Ensino Médio. Page 187. Brasília: Secretariat of Secondary Education and Technology.

BRAZIL (Country) Secretariat for Secondary and Technological Education - Ministry of Education and Culture. PCN + Ensino Médio: Orientações educacionais complementares aos Parâmetros Curriculares Nacionais. Brasília: MEC/SEMTEC, 2002.

BRITO. Ronilson Lopes. Citizenship Education in Chemistry Teaching. Monograph (Undergraduate Degree) - Full Degree Course in Chemistry. Federal Technological Education Centre of Maranhão. São Luís, Maranhão, 2008.

CAJAS, F. La alfabetización científica y tecnológica: la transposición didática del conocimiento tecnológico. Ensenanza de las Ciencias, v.19, n.2, 2001.

CARRARA Jr., Ernesto, MEIRELLES, Helio. The chemical industry and the development of Brazil - 1500 - 1889. Volume I and II - São Paulo: Metalivros, 1996.

CARVALHO, A. M. P. GIL-PÉREZ, D. Formação de professores de ciências: tendências e inovações. São Paulo, Ed.Cortez, 1998 (Coleção questões da nossa época).

CHASSOT, A. Alfabetização Cientifica: questões e desafios para a educação. 2ª Edition. Ijuí: Editora Unijuí, 2001 (Chemistry Education Collection).

______ . Catalysing transformations in education. Ijuí: Ed. Unijuí, 1993. p.174.

DELIZOICOV, D.; ANGOTTI, J. A and PERNAMBUCO, M. M. Ensino de Ciências:

Fundamentals and Methods. São Paulo: Cortez, 2002

DELORS, Jacques et al. Education: a treasure to be discovered; report for UNESCO of the International Commission on Education for the 21st Century. São Paulo; UNESCO/Cortez, 2003.

DEMO, Pedro. Desafios modernos da educação. Petrópolis: Vozes, 1993.

FREIRE, Paulo; SHOR, Ira. Fear and Daring: The Daily Life of a Teacher. Translation by Adriana Lopez, Rio de Janeiro. 11ª ed. Paz e Terra: 1986.

FREIRE, Paulo. Education and Change. Translated by Moacir Gadotti and Lilian Lopes Martin. Rio de Janeiro: Paz e Terra, 15th Ed. 1979.

. Pedagogy of the Oppressed. 24. Ed. - Rio de Janeiro: Paz e Terra, 1970.

______ . Pedagogy of Autonomy: Knowledge necessary for educational practice. 9. ed. - São Paulo: Paz e Terra, 1996.

FILGUEIRAS, C. A. L Origens da Ciência no Brasil, Química Nova, vol. 13, n. 03, pág.222 - 229, 1990.

JIMÉNEZ LISO, M. R.; SANCHES GUADIX, M. A.; DE MANUEL, E. T. Química cotidiana para la alfabetización científica: realidad o utopía? Educación Química, v.13, n.4, 2002.

KINALSKI, Alvina Canal et. a.l. STUDY SITUATION: Transdisciplinary Proposal for the Sciences of Nature and their Technologies in Secondary Education at the Francisco de Assis Basic Education Centre. In: GALIAZZI, M. do C. et. al. (eds). Networked Curriculum Construction in Science Education: a commitment to research in the classroom. Ijuí: Editora Unijuí, 2007. p. 357 -373 - (Science education collection).

LAKATOS, Eva Maria, MARCONI, Marina de Andrade. Fundamentals of scientific methodology. 3rd ed. rev. and expanded. São Paulo: Atlas, 1991.

LIBÂNEO, José Carlos. Democratisation of the Public School: The Critical Pedagogy of Content. 11th ed. São Paulo: Edições Loyola, 1993.

LOPES, Alice C. Curriculum and epistemology. Ijuí: Ed. Unijuí, 2007. 232 p. LUTFI, M. Os Ferrados e Cromados: produção social e apropriação privada do conhecimento químico. Ijuí, Ed. UNIJUÍ: 1992.

_____ . Everyday Life and Chemistry Education. Ijuí, Ed. UNIJUÍ: 1988.

MALDANER, Otavio Aloisio; et al. Contextualised Curriculum in the Area of Natural Sciences and their Technologies: the Study Situation. In: Fundamentals and Proposals for Teaching Chemistry in Basic Education in Brazil. ZANON, L. B.; MALDANER, O. A. (eds). - Ijuí: Ed. Unijuí, 200. p. 61-2.

MASTERTON, William L., SLOWINSK, Emil J., STANITSKI, Conrad L. Principle of Chemistry. Trad. J.S. Peixoto - Rio de Janeiro: LTC, 1990.

MORAES, R. The production of chemical knowledge and the teaching of chemistry: movements between everyday knowledge and chemical knowledge. Round table at the XIV National Meeting on Chemistry Teaching, Curitiba, 2008.

PERRENOUD, P. Ten new competences for teaching: an invitation to travel. Porto Alegre: Artmed, 2000.

_____ . A pedagogia na escola das diferenças: fragmentos de uma sociologia do fracasso. 2.ed. Porto Alegre: Artmed, 2001.

PERRENOUD, P. et al. Training professional teachers: What strategies? What competences? 2. ed. rev. Porto Alegre: Artmed, 2001.

SANTOS, W. L. P.; MORTIMER, E. F. Concepções de Professores sobre Contextualização Social do Ensino de Química e ciências. In: Annual Meeting of the Brazilian Chemical Society, 22, 1999, Poços de Caldas, MG. Book of abstracts. São Paulo: Brazilian Chemical Society, 1999b.

SANTOS, Wildson Luiz Pereira dos; SCHNETZLER, Roseli Pacheco. Educação em Química: compromisso com a cidadania, 4 ed. Ijuí: Ed. Unijuí, 2010.

_____ . Chemistry education: a commitment to citizenship. 3. ed. Ijuí: Ed. Unijuí, 2003.

TRIVINÕS, August N. S. Introdução à pesquisa em Ciência Social: a pesquisa qualitativa em educação. São Paulo: altos, 1987.

VALADARES, Jorge (2001) - Constructivist and investigative strategies in the teaching of science - Conference given at the Meeting "The Teaching of Science within the New Programmes" at the Faculty of Engineering of the University of Porto, 2001.

XIMENES, Sérgio. Minidicionário Ediouro da Língua Portuguesa - 2ª Reform edition. São Paulo - Ediouro 2000.

WARTHA, E. J. E FALJONI-ALARIO, A. El concepto de contextualización presente em los libros texto de química brasilenos. Educación Química, v.16, n.2, 2005.

ZUCCO, C. Undergraduate chemistry: a new chemist for a new era. Química Nova, v. 28, Supplement, S11-S13, 2005.